Dieter Mende

EEZ Energy, Energy industry, Future energies

The job engine of the energy transition

The global starting position,
resist the siren songs,
the prospects.

If someone tells you that they can explain the energy transition to you within a few minutes, then you should be extremely skeptical.

Thinking at EU/federal/state level and acting locally is not a contradiction, but rather a dynamic energy policy.

Dieter Mende

Dieter Mende

EEZ Energy, Energy industry, Future energies

The job engine of the energy transition

The global starting position,
resist the siren songs,
the prospects.

Impressum

© 2024 **Dieter Mende**, EEZ Energie Energiewirtschaft Zukunftsenergien
www.eez-mende.de

Further contributors:

Antje Mende, LIKES Layout – Impuls – Konzept – Entwurf – Style;
Moderne Medien-, Text- und Bildberatung

Herstellung und Verlag: BoD – Books on Demand, Norderstedt

ISBN: 978-3-7583-3031-5

Table of contents

Prologue

Aspiration and reality

Pictures:
German Federal Minister Svenja Schulze in
conversation with author Dieter Mende;
EEZ Energie Energiewirtschaft Zukunftsenergien

Today, not many people remember that the energy transition could have begun with a statement made by the German SPD candidate for chancellor Willy Brandt in his election campaign speech on April 28, 1961.

However, the voters had decided otherwise. At the time, nobody could have guessed that this would be the start of environmental policy in Germany:
„The sky over the Ruhr region must turn blue again".

"Clean air, clean water and less noise must not remain paper demands," Willy Brandt had demanded. This demand was triggered by research findings that had already shown in 1961 that the increase in air and water pollution resulted in an increase in leukemia, cancer, rickets and blood count changes, even in children. Willy Brandt had noted with dismay that this joint task required a great deal of effort and that industry also had to provide support when it came to the health of millions of people. At that time, people had just come out of the reconstruction of Germany after the war and were experiencing the upswing, witnessing the increase in jobs and had misjudged Willy Brandt's impulse as a threat to the incipient prosperity. In this respect, Willy Brandt's call "The sky over the Ruhr must turn blue again!" was both courageous and important. In the early 1960s, almost eighty blast furnaces and a hundred power plants emitted so much coal dust and CO2 that in winter the snow usually fell gray to sooty black from the sky.

Understanding the energy transition

The energy transition is not trivial and, with hydrogen, has unprecedented flexibility for the energy transition by coupling the sectors of electricity, heat, gas and fuel products; Power-to-X.

You will experience probably the most exciting developments in the modern world with the challenges of today; on the one hand with a view to maintaining the security of energy supply for people, and on the other hand with a view to the many opportunities for future generations.

A book about the energy transition must also deal with structural change and is also a testimony to the great efforts and challenges that the regions make and overcome when the goals are clearly defined and when the regional identity with these goals is supported intercommunally.
A book about the energy transition is also a testimony to the great efforts and challenges that companies in the energy markets are making and overcoming when established structures are broken up and networked with new or complementary business models, when complementary services are created with the expanded energy products.

The energy transition is much more than just the increasing use of renewable energies!
The energy transition is a job engine.

The success of the energy transition therefore depends crucially on the start and speed of implementation of defined goals, which requires determination and regional identity with the emerging fields of action.

The Emscher-Lippe region deserves special mention here; this region is an ideal example of how much holistic regional identity the individual steps and paths in the energy transition require in order for target agreements to be implemented in a sustainable manner.

The history of the Emscher-Lippe region is also a success story of the people living there, who have helped to meet the challenges caused by structural change; the topic of human capital will be a recurring theme throughout the book.

There is an old saying: "It is easy to see new horizons when you are sitting on the shoulders of a giant."

The Emscher-Lippe region in the northern Ruhr area has also been hit hard by the closure of the mines and the resulting structural change. The Emscher-Lippe region has not been able to call for a giant to shoulder the region and carry it forward to new destinations; the challenges for the region are still numerous today. The regional nucleus h2herten with its expansion into supra-regional commitments with the h2-netzwerk-ruhr is exemplary.

If the energy transition succeeds in the demanding and densely populated Emscher Lippe region, then the energy transition can succeed everywhere! Climate change is already clearly visible and the consequences are coming faster than feared.
One of the "three words of the year 2019" was climate youth!
The people who have been campaigning for the energy transition for years, not only the youth with Fridays For Future, for example, but also the many organizations, could not understand the actions of the German Federal Minister of Economics Peter Altmaier (CDU). In the course of the 2021 federal election campaign, the CDU's election program caused great resentment among the people; this with regard to the energy transition with climate change and environmental protection due to the lack of concrete statements on how the energy transition should be implemented.

On the one hand, the Federal Minister for Economic Affairs, Peter Altmaier, described renewable hydrogen as an important pillar for the success of the energy transition. On the other hand, the same minister prevented the expansion of wind energy by imposing such stringent requirements that many of the existing wind turbines could no longer be erected in this location today.

If the energy transition is to succeed, there is no alternative to the expansion of renewable energies. If the targets set with regard to phasing out coal are to succeed, politicians must ensure that the conflicting political decisions are changed immediately to ensure the success of the energy transition. These currently prevailing discrepancies are the subject of the admonishing people in Germany.

People in the EU have been aware of the fact that world peace is directly dependent on the energy transition, and not just since the latest climate developments. The effects of climate change hit people in poorer countries particularly hard, as well as people in Africa.
If the living conditions of the poorest continue to deteriorate and people move to more temperate regions of the world, this will have a significant impact on world peace.

People in the EU have long been unable to understand and now no longer accept the contradictory statements on the climate reports by those who are lobbying for the current coal policy; the slowing down of the energy transition had a major impact on the outcome of the 2021 federal elections.

Those who are involved in well-paid lobbying for the current coal policy with deliberately incomplete impulses, with the aim of unsettling the population, are now faced with the decision as to whether their actions are justifiable, whether greed for wealth is weighted higher by lobbying than responsibility for future generations, at the latest in view of the current climate reports.

The fact that insurance companies are also investing in the technologies and infrastructure of the energy transition is justified by the need to avoid increasing losses, such as those caused by more extreme weather due to climate change.

Citizens' wind cooperatives are an example of how the energy transition in conjunction with citizen participation not only has a positive effect on reducing climate change, but also offers a very interesting investment opportunity for people, in which everyone can decide for themselves how much money they want to invest in citizen participation.

The energy transition has become a job engine in Germany; technological expertise and infrastructure know-how are in demand worldwide.

Antonio Guterres (UN Secretary-General) was prompted to call on politicians to act quickly in view of the results of the World Climate Report of August 2021.

Svenja Schulze (German Federal Minister; SPD) warns in view of the results of the World Climate Report of August 2021 with the words: "The planet is in mortal danger".

The global starting position

Global opportunities and the image of the energy transition

In the course of the book, it is shown that and how the energy carrier hydrogen creates the bracket that encompasses the energy transition together with climate change, environmental protection, energy supply security ... all the way to world peace.

All the way to world peace?
How is that possible?

On the one hand, the energy transition has the goal of increasingly avoiding the burning of fossil fuels.
Hopefully, wars over fossil fuels will soon be a thing of the past.
The global focus on wind and solar energy is not as concentrated on regionally productive parts of the world as it is on the availability of fossil fuels such as oil and gas, so that no new geopolitical power factors will emerge.

Renewable energies such as wind, solar, geothermal, hydropower, biogas and the entire spectrum of renewable energy production have the potential to make the world a more peaceful place.

On the other hand, halting the effects of climate change is also important with regard to countries in Africa, for example, where life is already difficult for the people living there due to poor soil and a lack of water.

If climate change makes life impossible for people in these regions and millions of people make their way to the more temperate zones of the world as a result, then world peace will be threatened.

China has also recognized these correlations; China less with a view to a possible migration of peoples triggered by climate change, China rather with a view to Africa as a future location, which offers enormous opportunities for expansion with the potential of renewable energy generation from the sun. China is already lending a lot of money to African countries; money that the African countries will probably not always be able to pay back in view of the contracts, meaning that some of the infrastructure development promoted by China, such as the ports in Africa, may become the property of China.
It is currently clear that China is building up geopolitical power by maximizing its know-how with a view to future technologies and by striving to gain a share of global infrastructure.

However, it is unlikely that a country with the economic potential of the energy transition can achieve such great global power that it can virtually paralyze the world with an energy embargo, as OPEC did with the oil embargo in 1973.
If a country should allow a trade war to break out, the renewable energy plants that have already been installed will continue to generate energy.
The EU had already emphasized in the millennium year 2000 that the time had come for a uniform European energy policy.

Africa is not the only country to be mentioned when talking about renewable energy production as an export commodity. The focus is also shifting to countries such as Iceland and Norway, which are increasingly being referred to as the Kuwait of the day after tomorrow in analogy to their wealth of crude oil.
Iceland generates a large part of its energy from geothermal energy; the geysers of Iceland demonstrate the enormous availability of energy. Iceland needs only a fraction of the geothermal energy for its own energy requirements. Norway generates a lot of energy from the wind; the average annual wind speed in Norway is 38 km/h.

In Russia, the USA, Europe, Africa and Australia, the potential for renewable energy generation is known all over the world.
The hydrogen energy storage system is the global link that connects the globally distributed renewable energy sources with technological know-how and infrastructure expertise.
Furthermore, the energy transition has the potential to ensure that prosperity in the world, which is essentially linked to access to energy, can be distributed more fairly.
Energy supply and healthcare - how do they go together?
With the following link, the EU documents the increasing influence of mankind on the earth's climate and temperature, triggered by the use of fossil fuels, deforestation and livestock farming.

With this link, the EU documents the causes of climate change and also the consequences of climate change and, in addition to the dangers for the plant and animal world, also names the dangers for human health.

The EU's climate policy:
https://ec.europa.eu/clima/change/causes_de

In issue 3 in September 2013, the UMID Environment and People Information Service presented the main topic of the energy transition and health with the participation of the Federal Office for Radiation Protection, the Federal Institute for Risk Assessment, the Robert Koch Institute and the Federal Environment Agency.

It was emphasized that the measures of the energy transition can help to reduce the burden of disease. It was also pointed out that in the past the cross-cutting issues have not been dealt with in the same way, so that the cross-sectoral results, such as the recognized health aspects, cannot be shown to be compatible with the objectives of the energy transition.

This is precisely the reason why citizens have still not fully accepted the energy transition.
As a result, the energy transition is actually a renewable energy mix combined with a storage mix, which is completed with the sector mix:
electric power with gas, with heat and with user coupling mobile and stationary.

Such a complex combination without a holistic view has not been able to create acceptance in the past with regard to the skeptics of the energy transition; such a complex combination requires a high degree of practical demonstrations for citizens.

Decarbonization, the renunciation of burning fossil fuels such as coal, oil and natural gas, reduces air pollutants; this has been easily explained to citizens. The policy was able to trigger a great deal of acceptance because burning fossil fuels is a major cause of lung disease.

However, there has already been a lack of communication regarding land conflicts with regard to cultivation in the agricultural sector; the production of bioenergy from renewable raw materials such as energy crops (maize, etc.), such as wood, would be in competition with land for the food industry and would also endanger the areas of nature conservation areas.

Generating the same amount of electricity from bioenergy actually requires a hundred times more land than generating this energy with solar panels.

Wind turbines do not create any relevant land conflicts, e.g. with regard to agriculture.

The image of the energy transition has improved in recent years.

While some twenty years ago, the lobby for fossil fuels such as coal, oil and natural gas argued that renewable energy could not provide a base load due to the volatile, fluctuating generation of electricity, conventional power plants were presented as having no alternative. This argument has been invalidated with hydrogen energy storage.

Around ten years ago, the lobby for fossil fuels such as coal, oil and natural gas argued that even with hydrogen storage, renewable energy would be far from sufficient to cover the high energy demand in the EU, the grid centrality of conventional power plants was emphasized and unjustified grid problems were cited with decentralized renewable energy generation.

Projects such as the hydrogen user center h2herten have shown that renewable energy generation can be used to manage systems that guarantee a secure supply as an isolated solution, while at the same time having a stabilizing effect on the grid with parallel operation and intelligent grid management; the virtual power plant.

Just seven years ago, the lobby representatives of the fossil fuels coal, oil and natural gas argued that the challenges of the energy transition were not yet technically mature due to the lack of a serious and holistic discussion on the energy transition.
The fossil fuel lobby countered this by deliberately posing incomplete questions to create uncertainty; positions that we encounter more often throughout the book with regard to the subject areas and the holistic view from energy generation to applications.

We often heard it said that we first have to find out whether electromobility with batteries or electromobility with fuel cells will prevail. It is true that in electromobility, vehicles with batteries and vehicles with hydrogen/fuel cells complement each other, just as vehicles with petrol engines and vehicles with diesel engines currently complement each other in combustion mobility.

It has often been said that security of supply with electrical energy can only be possible with grid centrality. It is true that renewably generated energy with hydrogen not only enables decentralized stand-alone solutions, but also enables grid-parallel systems that can be coupled with each other through intelligent grid management to form a virtual power plant.

It has often been said, for example, that Germany is an energy importing country and that energy in the form of hydrogen will also have to be sourced from abroad in the future because the country cannot provide enough renewable energy. It is true that hydrogen from abroad with its transportation infrastructure is considerably more expensive, that with the expansion of renewable energy generation through modern wind turbines in conjunction with the generation of electrical energy through modern solar plants, a considerable proportion of hydrogen production can take place in Germany and can have a price-reducing effect for energy consumers.

Every single kilowatt hour of renewable energy generated locally/regionally in Germany contributes to an important reduction in energy costs by avoiding global energy transportation routes.

The ancillary costs for living, right up to the production costs in the economy, are linked to the price trends for energy.

Acceptance of the energy transition has reached all sections of the population with regard to its necessity.

How both private interests and industrial goals, as well as municipal (regional) opportunities can be shaped with regard to investments, is currently developing potential.
Where the public sector is leading the way and the energy transition can be experienced and experienced, sustainable regional expertise is already evident today.

The creation of cooperative structures between municipalities and regions has shown that the current transition of hydrogen technologies from research and development to the market has created numerous opportunities along the value chains, starting with energy generation, energy storage, energy distribution and energy supply, through to applications - stationary, portable and/or mobile.

Electromobility, digitalization, climate change, environmental protection, structural change ... these complementary topics must no longer be considered separately.

The success of the European economy in international competition depends crucially on the start and speed of transformation processes in all company structures, which requires the willingness to cooperate and the transformation of company training.

Germany, for example, has internationally recognized expertise in areas such as hydrogen and fuel cell technology, wind energy, photovoltaics and biomass.
Industrialized countries around the world, and increasingly also the USA, China, India and Brazil, are making use of our energy expertise.

Small and medium-sized enterprises (SMEs) have taken on an increasingly strong role in the development and production of future-oriented energy technologies with the corresponding service offerings and services.

The excellent qualification of skilled workers, the human capital, is another key to the success story in the German state of North Rhine-Westphalia.

The question of why the implementation of the energy transition did not start a few years ago makes it important to take a look at previous processes in the economy.
In the past, the results of basic research and materials research, for example, tended to be taken up by innovative small and medium-sized enterprises (SMEs).
In the past, the results of basic research and, for example, materials research were also taken up by
by company founders who use the results to build up a business segment.

In the past, the resulting entrepreneurial results or products were bought by large companies; in some cases, SMEs were taken over by large companies, e.g. with regard to patents.
In the past, large companies were referred to as lead companies because they were able to exert a significant influence on the markets.

With a view to the energy transition and with a holistic view from the source, from renewable energy generation, to the sinks, to the applications, a whole "bouquet" of entrepreneurial opportunities has emerged.

Although the diversity of opportunities associated with the energy transition has been presented with its mode of action, the diversity of opportunities associated with the energy transition has not been presented in the time-limited considerations of the balance sheets of large companies, whose shareholders focus above all on maximizing the profits of existing business areas.

This not only created the situation that the energy transition did not fit into the target agreements of large companies, which usually run for one or two years, it also created the situation that, from the shareholders' point of view, the energy transition created an economic sector that could compete with existing business areas.
This is due to their product-oriented approach, which has often not allowed for a holistic view.

If this was the case in the past, how did the first energy transitions from burning wood and coal to burning oil and natural gas succeed?
... From oil and natural gas combustion to nuclear energy?

About the niche markets.
Wherever an optimization has been found in the processes, e.g. to simplify workflows or to achieve the desired results, inventions and developments have been integrated into existing processes; initially in a very specific application, in a market niche.

Niche markets are not temporary!
Niche markets are the first entries into real markets.

Even the most successful developments on the market have their origins in a special segment with an initially incomplete value chain.
Expanded market models or new and complementary market models emerge at the tangents of niche markets.

In the past, regions have been able to successfully create numerous jobs by recognizing and establishing entrepreneurial activities in niche markets.

Recognizing a market niche does not necessarily result from a supply or market gap.

Recognizing a market niche can also result, for example, from the attractiveness for consumers (mainstream) or, as in the case of the energy transition, from an environmental policy necessity.

But why is it now the globally active energy supply companies that are starting the current energy transition with hydrogen energy storage?

Because a sustainable infrastructure in the energy transition couples the existing electricity, gas and heating sectors with smart grids, creating a market environment that allows the existing market mechanisms to transition into bridging technologies that pave the way for future technologies.

<u>The energy carrier/storage medium hydrogen is a key pillar here.</u>

A closer look at the technical potential of power to heat, power to fuels, power to gas and power to chemicals with a view to coupling the electricity, gas and heat sectors in smart grids reveals material flows at whose tangents further potentials arise. Hydrogen energy storage is playing an increasingly important role in more and more technical potentials; if produced from renewable energy, it is also environmentally friendly and climate-neutral.

In order to recognize and understand these connections, it is important to look back into the past, as this reveals the trigger for the past lack of dynamism.
Is it really so easy to point the finger at large companies in the energy industry or at vehicle manufacturers?
No.

The goal of a company should be to maximize profits. It is perfectly legitimate for a company to protect this goal in competition with other companies and other products. Competition is not a threat in the markets; competition is often the trigger for innovation.
Why the current energy transition is only now being communicated as an innovation and no longer as a threat to the markets is not solely due to environmental policy necessity.

The reason for the past lack of dynamism in the energy transition is not just that a whole "bouquet of flowers" of entrepreneurial opportunities with regard to the energy transition and the holistic view from renewable energy generation to applications would have been confusing.

The reason for the past lack of dynamism in the energy transition lies in the combination of the diverse opportunities identified by a diverse corporate landscape, which were focused on their own business areas that had previously been considered separately.

The famous "thinking outside the box" quickly touched on other business areas, so that further communication did not take place.

To illustrate this list:

The key technology for the success of the energy transition is, on the one hand, hydrogen technology with the storage and provision of energy.
On the other hand, the key technology for the success of the energy transition is energy conversion, which also includes fuel cell and electrolysis technologies.
So far, this is technically understandable.

In the past, the orientation of the companies was different with regard to their entrepreneurial responsibilities.

The key companies for the success of the energy transition are numerous:
+ the energy suppliers,
+ the chemical industry
+ plant engineering,
+ mechanical engineering,
+ the automotive industry,
+ and many more.

One would think that if an entrepreneurial opportunity arises with the energy transition through the energy storage of hydrogen and through the energy converters fuel cell and electrolysis, that an overall interest of the companies would "naturally" arise from this; that the chemical industry with a view to hydrogen would enter into a dialog with the plant engineering industry with a view to fuel cells and electrolysis.
This focus also shows once again:Thinking EU/federal/state-wide and acting locally is not a contradiction, but rather dynamic energy policy.

It's the same here as in the past:
one:r simply has to do it once and get started.

The key to regional development can be found in the competence centers of the federal states.

The key to implementing technological know-how lies in human capital with additional qualifications.

The key to economic development lies in constructive networking, also via advisory boards.

The key to strengthening development processes lies in proactive communication with a local identity.

The key to open structural development can be found in the change processes that accompany the transformation from the coal and steel industry to the industry of the future, for example.

The key to promoting competence fields is the establishment of innovative working teams consisting of employers, trade unionists and labor market experts.
Players with the necessary "inner drive" for change.

The key to implementation is the regional programs and regional projects.

The keys listed here were in the past, and still are today, tried and tested processes that trigger each other and work together successfully towards the goal; figuratively speaking, like a row of dominoes.
However, if the first impulse does not occur, the tried and tested series of keys mentioned cannot function.

In view of the complex interrelationships mentioned here, it is therefore also evident in the energy transition that once the opportunities have been recognized, one simply takes the initiative.

The regional nucleus in the Emscher-Lippe energy region is h2herten, which has quickly gained supra-regional importance and to this day has one of the rare centers in the EU that successfully demonstrates THAT and HOW the energy transition works; the hydrogen user center h2herten.
The success of the hydrogen user center h2herten is confirmed not only by the international flow of visitors, but also by the international cooperation with companies from Japan, Canada and the USA.

Bild: Panoramic view of the Hoheward spoil tip. The hydrogen user center h2herten is on the right in the picture, in the background you can see the preserved existing buildings of the former Auf Ewald colliery. Looking from this location, the hydrogen filling station is obscured by the hydrogen user center h2herten; Dieter Mende, EEZ Energie Energiewirtschaft Zukunftsenergien

The outlined timeline shows the supra-regional radiant dynamic, triggered by the h2herten nucleus, which led to the HyExpert Kreis Recklinghausen, which triggered the founding of the h2-netzwerk-ruhr with the hydrogen energy storage system:

1996   Identification of hydrogen as a valuable energy carrier contribution with an increasingly renewable energy production.
1999   Identifying the possibilities and opportunities of hydrogen energy storage in the energy infrastructure.
2002   Identification of regional and supra-regional service providers.
2003   Establishment and expansion of the resulting networks.
2006   The feasibility study commissioned by the city of Herten reflects the identified results more than positively.
2006   Bundling hydrogen and fuel cell projects in the northern Ruhr region with Herten.
2007   Advisory board established initially for the region, then expanded.
2008   Foundation of the h2-network-ruhr.
2009   Opening of the hydrogen user center h2herten.
2009   Competing fuels and technologies have made the justifiably forward-looking development of fuel cell and hydrogen technologies shine beyond the region.
2010   The constant involvement of other service providers beyond the borders of North Rhine-Westphalia.

The foreseeable penetration of hydrogen technologies in the most diverse markets opens up the expansion of traditional production processes and also the expansion of established industrial structures.

The development of an action plan with the aim of covering the entire value chain in the energy transition is the result of proactive teamwork in Herten, which has taken a holistic view of fuel cell and hydrogen technologies, starting with product development and extending to applications.

It is no coincidence that the HyServ-Center of the NRW/EU project HyChain-Minitrans, which was approved by the EU Commission on January 31, 2006, was established in Herten. At the time, HyChain-Minitrans was the largest EU joint project for the market launch of small and light vehicles with fuel cell drives. Light vehicles and mini-buses had started demonstration operation for the identification of further approaches; the aim of the HyChain-Minitrans project was the introduction of hydrogen as an alternative fuel on the basis of innovative fuel cell vehicles in the low power range up to 10 kW.

Various vehicle concepts were developed in the project; not in series, but rather through the conversion of existing vehicle models, which were used in the four regions:
+ France, in the regions of - Grenoble,
+ Spain, in the Soria/León region,
+ Italy, in the Modena region,
+ Germany, in the Emscher-Lippe region.

The various vehicle concepts were used and tested over a three-year period.
The hydrogen competence center H2Herten served as the location for the project management and also served as a qualification and information center.
The coordination of the three-year test phase of the vehicles, the development and coordination of training and support services and the development of marketing had generated valuable results.

The results have made it clear how important qualified employees are if the results developed are to be implemented; there was no special training for fuel cell technologies.

For the companies involved in the h2herten nucleus, workers in radio and television technology were identified for the additional qualification, as they were used to assembling delicate components and were also experienced in handling high-quality components.

This was quickly followed by the realization that the energy transition must also find and offer solutions in the area of additional qualifications. The energy transition innovator not only generates infrastructure capital, the energy transition job engine also generates human capital.

With a view to the challenges of creating human capital, the educational aspects were not neglected in Herten. h2herten established a dialog with Forschungszentrum Jülich, Institute for Materials and Processes in Energy Technology (IWV), Energy Process Engineering (IWV 3) at a very early stage from 2004 onwards.

In addition to the research and development activities relating to the fuel cell, IWV-3 has devoted considerable effort to the aspects of career guidance, both for school leavers and for courses, as well as for students and also for further training measures for engineers, technicians and craftsmen.

With regard to the research and development of fuel cells, the focus has been on the motivated introduction to technical tasks in the laboratory and workshop area; the science and application-oriented transfer of knowledge to universities and universities of applied sciences as well as the well-founded introduction of trainers to the teaching of a new technology has been the focus of the work.

A future-oriented contribution for basic, advanced and further training is focused on different levels of requirements in the sense of a low-delay introduction and a desired establishment of the fuel cell in the markets of energy applications.

The author Dieter Mende has provided an excellent overview in this article:

+	Background
+	Projects and actors
+	Analyses
+		Target groups
+		Craft
+		industry
+		Secondary school and university
+		public
+		Timeline
+		Contents of training and further education
+		Preparation of users
+ Implementation
+		Transfer level: provider - intermediary
+		Transfer level: provider - user
+ Assessment of qualification requirements
+ Summary
+ Outlook

At this point the circle closes:

In summary, the following can be listed:
+ the potential links through competence networks with the economy result from the bundling of innovative steering groups.
  + How constructively is communication constructively promoted?
  + how are partners recruited through regional competence?
+ Are lead companies really an innovative development potential?
  + with regard to the development/expansion of infrastructure?
  + with a view to technology development?
  + with regard to the competence networks in terms of applications?
  + with a view to industrial policy development?
  + with a view to structural policy development?
  + with a view to research and development?
  + with a view to cooperation and projects?
  + with a view to market developments?
  + with a view to the qualification of professional groups?
  + ... or are profit-maximizing companies blocking innovation?
+ Industrial clusters promise development potential for hydrogen and fuel cell technologies.
  + How do regional competence centers approach the the topic?
  + Which regional programs / regional projects result from this?

Looking at the initial question of potential, current technical possibilities, research and development, efforts to create a sustainable infrastructure, project developments and expectations in terms of the timeline until the introduction of the new technologies in the energy transition generates the question of the market ramp-up of hydrogen and fuel cell technologies.

The development work on fuel cells has already matured into a real competitive factor. Fuel cells represent an economic opportunity to optimize, secure and expand high-tech jobs and corporate success in the German economy.

The success of German industry in international competition depends crucially on the start and speed of the transformation processes in all company structures, which requires the willingness to cooperate and the transformation of company training.

Innovative steering groups must clearly emphasize in their external presentation to industry that hydrogen technology is wanted regionally.
This requires regional information and demonstration work. Only if cluster management is positively recognizable can companies' doubts about investing in a location be dispelled.

The current goal must be to bundle the opportunities in the technology field of fuel cell and hydrogen technology to create sustainable jobs.

The starting position of a location in the field of fuel cell and hydrogen technologies is not characterized by:

+ nationally renowned scientific institutions,
+ well-known large companies with a focus on application,
+ industrial settlement with a focus on production,
+ industrial settlement with a focus on plant construction,
+ Supply infrastructure with a focus on energy sources,
+ concentration of manufacturers of hydrogen and other fuels,
+ Qualification work in the education network with a focus on human capital.

With regard to the claim to an industrial and commercial leadership role of individual regional competence centers in Germany, they would be well advised to rethink their regional presence and move towards complementary networking that incorporates the entire potential grid of the energy transition.

The image of the energy transition in the field of fuel cell and hydrogen technologies varies from region to region and depends on regional networking.
The development of a region can only be successfully managed if the entire industrial and scientific environment of the country is included in the project work.

This is not a natural process, but rather a pro-active project management process implemented by a cross-sector network management.

With the wave of euphoria surrounding hydrogen as an energy source in the years 2000 to 2002, some location management companies named their development work "lead region" and are currently running the risk of losing recognition as an "announcement region", as the entire potential grid of the energy transition has not been included.

Location managements that have established and expanded cluster management without the industrial constraints of a lead company in accordance with the objectives and priorities of integrated state, economic and labor policy now enjoy international recognition and a leading position throughout Europe that was unimagined at the beginning.

Competence regions whose location management and development work are linked to the industrial orientation of a lead company (e.g. an energy supply company or a car manufacturer) cannot operate without industrial constraints. With a view to the development and expansion of infrastructure, technology development, competence networks with regard to application, industrial and structural policy development, research and development, the scope of cooperation in the projects, market development and the qualification of professional groups, the question must be allowed as to what development potential can still be pursued if the lead company discontinues project work?

For a long time, politicians were actually convinced that it would be the leading companies in the energy industry and mobility sector that would help the development of fuel cell and hydrogen technologies to market maturity.
But these companies are more likely not.

How a fuel cell works technically is well known.
Gas processing is no longer a technical novelty. So where can we find the way to advance development work on fuel cell and hydrogen technology?
In the past, the system interfaces may have caused problems, which is why research is the extended workbench.

If the companies and scientific institutions are brought together in the joint projects and demonstration projects, innovation is created.
This is a development that the energy transition has lacked for the reasons mentioned above for successful and holistic implementation.

The success of the energy transition is not necessarily linked to the technical simultaneity of fuel cell and hydrogen technologies. The industry's clearly recognizable focus in the energy transition is based on fundamental technical reliability, customer orientation, market proximity, system diversity, market relevance, economic efficiency and environmental compatibility.

The focus on development work for the success of the energy transition includes the sub-disciplines of system adaptation, service life of technical components, energy density, cost reduction, standardized component interfaces, modular system design, process dynamics, system dynamics, load dynamics, security of supply and, again, environmental compatibility.

The energy transition as a job engine

The energy transition job engine and human capital.

The energy turnaround as a job engine, the technological know-how and the infrastructure know-how that comes with the energy turnaround is in global demand.
Pictures:
Dieter Mende, EEZ Energie Energiewirtschaft Zukunftsenergien
Photo staged with advice from
Antje Mende, LIKES Layout Impuls Konzept Entwurf Style

The author Dieter Mende has taken a holistic approach to the development and expansion of sustainable infrastructure with hydrogen energy storage;
Networks work in and with networks (focus: EU),
cluster management (focus: Germany):

+ New structures, partnerships, customer bases recognize, plan and implement
    + Design
    + Development
    + Structure
    + Implementation
    + Support
+ Old structures, partnerships, customer bases reach, evaluate and optimize
    + Maintenance
    + Modernization
    + Remodeling
    + Implementation
    + Support
+ Force fields: Observation, sounding out, updating
    + Goals:
        + Target groups
        + Detailed objectives
    + Organizations:
        + Participants
        + Structures
    + Management:
        + Steering groups
        + Innovative working teams
    + Tasks
    + Competencies

+	Structure plan:
    +	Target estimation
+	Fields of forces:
    +	Actors: active: pro
    +	Actors: active: against
    +	Actors: passive
+	References to other projects:
    +	Consequences
    +	Structure plan
    +	networks
+	Market trends:
    +	Assessment within the EU:
        +	Markets in the next 10 years
        +	Medium-term to 2035
        +	Long-term until 2050
+	Project management (focus: NRW and Germany)
    +	Development of the markets

The intergroup statements by politicians already show that hydrogen will definitely become established as an energy storage medium to long term.

Hydrogen is completely independent of the mineral oil industry. The perspective for the future is the direct use of hydrogen. The summary of the core statements shows that there is a great deal of development potential in North Rhine-Westphalia and that there is also a great deal of well-founded know-how in the companies.

A look at the search engines on the Internet immediately brings up the explanation that human capital is the entirety of the economically exploitable skills of employees, the knowledge and behavior of individuals or groups of people in a company.

If we look a little further and take a closer look, it becomes clear why the term human capital quite rightly includes the word capital.
On the one hand, a closer look makes it clear why human capital in particular is important for the success of the energy transition; on the other hand, a closer look makes it clear that the energy transition is a job engine that can actually increase the employment figures of the workforce exponentially.

With hydrogen energy storage in the energy transition, a development is emerging in the markets similar to the exponential growth we have already seen in telecommunications, robotics and microtechnologies.

The growth effect of exponentially growing markets is not based solely on the willingness of companies and/or service providers to invest; the growth effect of exponentially growing markets is based at least to the same extent on the "inner drive" of employees, who drive the company forward with its entrepreneurial goals.

This is complemented by the realization that success can be really fun if everyone involved can/is allowed to enjoy it fairly.

Appreciation of the innovative, motivated employees by the management is shown through fair financial participation, and perhaps also through profit distributions.

In the past, motivated employees have often been the catalyst for operational success.

The energy transition has already shown with its start what dynamics can become possible. The energy transition, climate change and environmental protection are inextricably linked and their implementation will trigger a requirement that not only means flexibility for the users of the future, but also flexibility for employees in companies with regard to upcoming additional qualifications. The energy transition will also transform one or two jobs into adapted employment; the energy transition is already creating additional jobs.

Without the "inner drive" of employees in companies, it will not be possible to achieve the goals that have been set. An increasingly cooperative management style on the part of superiors has become apparent in companies that are already affected by the implementation of the energy transition.

Materials research was and is an essential part of the development of new technologies. Seals, hoses, valves ... an unbelievable number of tiny components are installed, are integrated into the technical systems and determine the processes from operational procedures to production; determine the designs and processes of the programmed machines, have a significant influence on the efficiency of energy generation.
Rigid career paths, as we may remember them from years gone by, prevent innovative commitment in companies because employees perceive rigid career paths as a restrictive corset and because rigid hierarchies in companies also greatly reduce the flow of information.

The wealth of ideas, enthusiasm and motivation cannot be studied; the wealth of ideas, enthusiasm and motivation are the result of an interaction between employees and company management.
Stronger networking between employees, their interaction with each other, creates a high potential for innovative impulses.

It is perhaps now becoming a little clearer that human capital is not simply the entirety of the economically exploitable skills of employees, the knowledge and behavior of individuals or groups of people in a company.
We will come back to this point in the course of the chapter, thus closing the circle.

Resource efficiency and market trends justifiably show that the desire for an energy transition can also become a reality.

With regard to energy consumption, resource efficiency is above all also the endeavor to reduce energy consumption, e.g. with the same work performance of machines, both mobile and stationary.
Resource efficiency can be supported by a more conscious, economical use of energy, which includes not only electrical power but also heat. However, resource efficiency is also the endeavor to use fewer raw materials.

Prosperity and convenience are not at all contradictory to the desired reduction in the consumption of raw materials; on the contrary, as the past has shown us, innovation has very often led to greater resource efficiency and greater convenience at the same time.

Compared to classic cars, modern cars show that resource efficiency and convenience go very well together.

Reliability, convenience, prosperity, sometimes just the mainstream, are triggers for market trends.
What is currently influencing market trends is the recognition of the impact of climate change.

The recovery of raw materials, recycling, is becoming very important with the success of the energy transition.
The fact that cheap products have to be purchased more frequently due to a lack of material quality and are ultimately more expensive than high-quality products, and that cheap products result in more waste, has been recognized by the public.

People are increasingly seeing "cheap is cool" as an undesirable development; with the original launch of the Saturn chain's advertising slogan in 2002, a trend had developed that had also brought a great deal of attention to the "one-euro stores".
What triggered this trend in the background was the displacement of high-quality goods in the stores and, as a result, goods were produced cheaply, e.g. in China. The declining appreciation of value had triggered an effort in companies to offer equally cheaper goods, which immediately triggered a displacement of jobs by low-wage earners.
Companies often saw their employees as a cost factor rather than as human capital; this imbalance meant that employees were unable to provide management with any innovative impetus.

Where low-cost appliances are preferred, there can be no question of energy-saving use; the components installed in low-cost appliances are not designed for energy efficiency.

With the energy transition, the focus on energy efficiency has also become very important. Value, durability and environmental compatibility are the driving requirements that have triggered a trend reversal in the energy markets.

Suddenly, skilled workers were increasingly in demand; companies increasingly recognize their employees as creative and innovative again: the return of human capital in the perception of entrepreneurs.

A trend reversal can also be seen at the online auction portal Ebay.

Until around 2011, the slogan "Three, two, one, mine" was a call for the acquisition of second-hand goods, for the acquisition of cheap goods, of flea market items, but since then Ebay has made a U-turn with a focus on offers with fixed prices from commercial traders.

Value, durability and environmental compatibility are often emphasized with regard to the characteristics of the goods.

China has gained a great deal of product know-how in the years of cheap products through production orders.

Value, durability and environmental compatibility are therefore in competition with cheap goods, e.g. from China.

With the environmental necessity of the energy transition, people have also developed an understanding of sustainability.

In general, it can be observed that the energy transition as a job engine has triggered a trend reversal on the labor market.

The wealth of ideas, enthusiasm and motivation of employees in the companies are valuable impulses for new innovative products and trigger a spirit of optimism in the European market.

Many new markets could emerge as a result of the energy transition, but the potential for expansion that can already be realized now is also of extraordinary interest to industry, business, service providers and trade.

In view of the range of current fossil fuels and the increasing need for action to reduce/avoid emissions, the energy industry and energy efficiency are a crucial platform for the important and right changes. The energy industry is probably one of the most interesting sectors of the 21st century.

The integration of the German economy into the European economy began with the coal and steel industry: the first European Coal and Steel Community Treaty (ECSC) was signed on April 18, 1951 with Belgium, the Federal Republic of Germany, France, Italy, Luxembourg and the Netherlands for a period of 50 years.

With its historically active industries, the Ruhr region in Germany was regarded as "the land of ten thousand fires."

In terms of electricity generation and consumption in the Federal Republic of Germany, North Rhine-Westphalia is still the No. 1 energy state today. With the innovation of the German energy industry, conventional energy sources such as oil and gas have established themselves on the markets alongside the coal industry since 1951.

The decisive role of an energy location, an energy region, is the ability to generate and distribute energy economically, efficiently, ecologically, sustainably and in an environmentally friendly manner.

Both an excellent environment with universities and research institutions, as well as the economic and technical expertise of the companies are the innovative basis; with regional economic development the innovative engine.
This was true for conventional energies in the past and is still true for renewable energies today.

Whenever an energy crisis occurs, people look to NRW with great expectation.
The gas crisis in Ukraine at the beginning of 2006 also made it clear how important renewable energies are in the energy mix for our energy supply security.
Renewable energies in the energy mix are increasingly gaining a foreign policy dimension and are a topical issue in the EU.

The EU is striving for a European energy policy; this in connection with the completion of the European internal market, with the climate protection targets that have been set and with a view to energy supply security.
These are the major challenges in the energy sector that require joint European solutions.

Hybrid technology was developed in Germany, was completely misjudged by the German automotive industry in terms of its importance and was not integrated into the systems. In Asia, this hybrid technology has been adopted by the automotive industry.

The Toyota Prius is the first series-produced hybrid vehicle to come onto the market in which a conventional combustion engine is combined with an electric drive and a modern battery system. Mistakes of this kind must not be repeated in view of the technical possibilities offered by the energy transition!

In addition to the expanded products, the energy transition job engine also brings many new products and services to the markets.

The strengths and weaknesses of a start-up company with hydrogen and fuel cell technologies must stand up to the scrutiny of a sustainable energy infrastructure in order to expand opportunities and to highlight promising collaborative projects as a unique selling point.
The developed profile of a start-up company is the recommendations for action and the definitions of meaningful contributions with the aim of developing and expanding the infrastructure.

In fact, as shown earlier in the book, it is not enough to try to bring together so-called leading companies in one location in the hope that this will lead to success with a view to making important contributions to the energy transition.
Conversely, the location requirements are recorded for the acquisition of project partners, for cluster formation.
The development of unique selling points contains the potential for success of a regional nucleus.

Along the value chain from fuel cell production to applications, including the requirements of the fuel infrastructure, niche products and special applications must be sustainably positioned in the early markets.

In addition, highlighting competencies in terms of service is of great importance in order to counteract initial skepticism.

Innovative identification procedures with a view to making meaningful contributions to the energy transition can generate a competitive advantage for companies.

Sustainable fields of action have prospects of success.
Both the technology and market potentials, as well as the expectations of the markets in terms of growth, are often not sustainable without integration into the country's/EU's energy transition potential grid.

What has so far prevented the energy industry from providing the energy markets with an environmentally friendly energy carrier in the form of hydrogen?

The fact that the oil industry's resources are running out is not a new message. The process of a fuel cell application is free of environmental emissions. The desired decarbonization depends on the fuel used. The operating heat of the fuel cell can be used and is generated without exhaust gases. The energy carrier hydrogen has a very high energy density.

Why has the energy industry so far neglected the potential success of hydrogen as a storage medium for renewable energy with regard to security of supply?

For many years, there has been no pressure on the industry to rethink.

+	Inventions are less common in large-scale industry and more common in small and medium-sized enterprises.
+	Hydrogen does not occur in isolation in nature.
+	The environmental balance of hydrogen from fossil fuels is poor, as hydrogen has often been a waste product of other production chains.
+	The dwindling resources of fossil fuels have received too little attention.
+	The greenhouse effect, climate change and its global consequences have been called into question by the coal lobby.
+	A complete consideration of the possibilities of environmentally friendly energy production was lacking.

The central question of what a sustainable energy supply can look like has also been brought to the attention of those who previously did not want to deal with this issue as a result of the environmental policy challenge.

The regional nuclei have dealt with the question of energy supply security in the current energy transition dialog, with the question of the technical possibilities and efforts regarding a sustainable energy infrastructure, with the question of the status and results of project developments, with the question of expectations on the timeline until the introduction of new energy sources, including hydrogen, and can present valuable results.

With the EKS Energy Complementary System, the hydrogen user center h2herten demonstrates the holistic view of the energy transition and shows that and how the energy transition can be led to success.

Decentralized energy supply is indeed a paradigm shift for energy supply companies.

In fact, however, the energy transition is not consistently characterized by a decentralized energy supply; rather, the systems can be operated both as stand-alone solutions and in parallel with the grid.

Coupled with smart grids, digitalization can create a virtual power plant. This provides further and complementary approaches for a sustainable infrastructure in the energy transition.
The price of natural gas is linked to the price of crude oil, so it remains to be seen how energy costs will develop in the coming years. At the moment, politicians are already sending signals that the price of petrol will move towards two euros per liter.

Almost all the media are currently reporting on hydrogen and fuel cells.
Opinions on how the energy transition can be successful are quite contradictory. Nevertheless, the hydrogen initiatives have already achieved one goal: the public has become aware of hydrogen and it is the energy source of the future.

There are two large hydrogen distribution networks in the Federal Republic of Germany that connect industrial producers and consumers.

For around 80 years, a 240 km long hydrogen pipeline, starting in the northern Ruhr area in Marl, has been supplying industry with hydrogen from the south to Cologne. The operator of the hydrogen pipeline is Air Liquide.

A second hydrogen network runs in the Leuna-Bitterfeld-Wolfen region. The operator is Linde AG.

Another hydrogen pipeline is to be built between Emsland and the Ruhr region.
The pipeline between Lingen and Gelsenkirchen should be operational by the end of 2022.

The Port of Rotterdam Authority has launched a feasibility study for the construction of a new hydrogen pipeline from Rotterdam in the Netherlands to Germany.

The future demand for hydrogen as an energy storage fuel with a view to the introduction of series production of fuel cells will enrich the entire energy market.

A statement that is often used as a reminder:

The Stone Age didn't end because there were no more stones, and the hydrogen industry won't only get started when there is no more oil.

The conversion of the energy supply to a hydrogen energy economy is often said to take around 50 years; the conversion from coal to oil is used as a benchmark. However, it should not be overlooked that the following situation is assumed when the industry talks about a transition:

+   Security of supply at any time and in as many places as possible
+   A nationwide infrastructure:
        +   The supplier has sufficient distribution options available,
        +   the customer can choose between the products at any location,
+   an economical energy price.

Probably the most crucial aspect of our dependence on fossil fuels is the point at which, taking into account various consumption trends, the production of fossil fuels will no longer be able to meet our energy needs. According to current estimates, this point has already been reached,

In Germany, for example, the energy demand for heat is often underestimated; in Germany, more than 50% of energy consumption is used to generate heat.
Hydrogen as an energy source can also offer technical solutions for heat with power-to-heat. A sustainable energy supply with the important security of supply relies on new technology options being ready for the market at an early stage and being both economical and environmentally friendly, with the aim of achieving climate neutrality.

Following the increased closure of mines, interest in mine gas as a potential additional energy source has grown steadily. Even after the end of mining, more than 120 million standard cubic meters of mine gas per year escapes unused into the atmosphere from old mines.
The resulting drop in gas pressure in turn leads to the release of adsorptively bound gas.
This effect is comparable to a champagne bottle, where once the cork is removed, the carbon dioxide bubbles until there is none left.
Throughout Germany, 1.5 to 1.7 billion standard cubic meters of mine gas are released from mines every year.

The methane content of mine gas is approx. 70%. The comparison of climate-relevant emissions in Germany, for example, makes it clear that the use of mine gas can also become important as a bridging technology, as a storage facility for hydrogen.

The statistics on climate-relevant methane emissions for the example in Germany, per year:
+   Mining: 568.359 [ton/year]
+   Livestock farming: 22.300 [ton/year]
+   Landfill: 183.295 [ton/year]
+   Wastewater treatment: 107.280 [ton/year]

The energy supply companies RWE and E.ON had already announced the reduction of excess power plant capacity in Germany in 2008; the European energy markets are monitoring these developments very closely.

Regardless of whether the shutdown of power plants is politically motivated or whether the shutdown of power plants is due to ageing, the existing infrastructure with regard to the grid and buildings has future potential.
There is a real opportunity for fuel cells in power plants with regard to fuel cell types with a very high operating temperature (hot modules), as this means that local heating customers could also continue to be supplied.

In order to secure the competitive advantages for fuel cell technology in Germany in view of the very high level of development, the issue of infrastructure for fuel cell vehicles must now be addressed more quickly.

Since 2007, companies in the automotive industry and energy companies have been working on a German "transport industry strategy" with the aim of making new hydrogen-powered vehicles ready for series production.

The four largest Japanese automotive companies alone are investing 700 million euros of their own funds per year in the series production of fuel cell drives.

This closes the circle; we come back to the topic of human capital:

The skilled workers required by production companies are available on the labor market in NRW due to the sharp decline in heavy industry.
The current challenge facing the economy is the provision of suitable training places and the appropriate retraining and further training of employees.

The course of the book has previously shown that the success of German industry in international competition depends crucially on the start and speed of the transformation processes in all company structures, which requires the willingness to cooperate and the transformation of company training.

Just twenty years ago, skepticism and the hesitant attitude of industry towards fuel cell technologies were attributed to a lack of willingness to take risks or innovate.

At best, environmental protection and the desire to establish their own technology in the potential grid of the incipient discussion on the energy transition prompted companies to participate in initial pilot projects. Today, competition is no longer just a strategy; the aim is to secure the future viability of companies with regard to their products and services.
Missed opportunities in the transition period could have shortened the path to the economy with hydrogen as an energy source!

Here is a significant example:

Investigations by the Hamburg electricity company HEW in 2007 had shown that 1 million cars with fuel cell drive could be operated in Germany with the hydrogen that can be obtained from primary and secondary power plant regulation; this assumes that the cars have a fuel consumption of 10l/100km, that the cars have an annual mileage of approx. 12,500 km.

By integrating the city buses running on regular routes (approx. 7,000 buses in Germany), another 3/4 million cars could be powered by hydrogen. This could be achieved with practically no significant increase in emissions from energy generation.
In the past, market acceptance has been decisive for the breakthrough of new technologies, and this is still the case today.

Today, as was shown in the course of the book, product quality is once again prevailing:
+	provided the price is reasonable for a private family household,
+	provided that the operation is economically viable.

The initially hesitant market entry of cell phones and notebooks followed a similar course to the market entry of fuel cells. With the integration of the technical possibilities offered by hydrogen energy storage and fuel cells as energy converters, the existing value chains are supplemented and new value chains are opened up.

It is the value chains that define the fields of action and not the other way around; this also applies holistically to the potential of the energy transition.

The prospects

Are you now very worried about future generations?
Justified, if the energy transition is still not started.

A reason to panic?
Please don't.

Those of us who deal with the challenges of the energy transition know how the energy transition can succeed and we are committed to it. The fact that our commitment has been recognized can be seen in the political statements that hydrogen as an energy storage system will help the energy transition to succeed.

The aim of this book is to take you into the dialog that the energy transition is important and right; this with the start of implementation NOW.

The book, which is expected to be published in the third quarter of 2022, will present solutions, show initial approaches, demonstrate the effects and present the energy transition to a broad target group in such a way that the deliberately incomplete and/or false statements of those who are against the start of the energy transition will no longer cause you any uncertainty.
I would be delighted if we could come to the basic understanding together that the energy transition, with all its exciting and promising sectors, is the decision for sustainable coexistence.

City as storage

Smart technologies open up excellent opportunities for building planning, urban development planning and regional district planning; not only to reduce emissions, but also in the integration of renewable energy storage.
The reduction of individual energy consumption in conjunction with coupled energy systems through to network-controlled energy management enables efficient local district solutions. Superordinate intelligent grids can combine these efficient district solutions up to "virtual power plants". Virtual power plant is the term for an interconnection of decentrally generated electricity, such as electricity from photovoltaic systems, wind power and biogas plants. Heat generation and distribution can also be interconnected, e.g. with combined heat and power plants.

City as climate savior

What is increasingly affecting the sealed surfaces of cities are the effects of more frequent and stronger heat waves and more frequent and stronger rainfall. Cities and regions can combine the capacities of the local economy with the capacities of the public sector and actively make meaningful contributions to saving the climate in the potential grid of the federal state.
Cities can combine public and private funds, cities know the regional demand and can therefore optimize their procurement processes.

The author

Dieter Mende

Homepage:
www.eez-mende.de

The book is written with the background of professional training in chemistry as well as in electrical engineering, supplemented with the background energy dialog EEZ energy energy industry future energies; I myself am the founder of the energy dialog EEZ on 05.07.1995.

My professional background is wide-ranging:

Commissioned since September 1997 with the central control technology for energy centers in the automation technology project office of a regional energy supply company; in November 2019, I moved to the electrical power grid construction project planning department.

Since February 2003, initially commissioned with the tasks of setting up and expanding the regional hydrogen nucleus h2herten, then commissioned with tasks of project and company acquisition, market communication and net-working in the team of the hydrogen user center h2herten.

In both cases, the employers noticed the very successful EEZ energy dialog, which led to the jobs.

My motivation for creating reports and books is, on the one hand, a passion for highlighting the opportunities and possibilities in the potential grid of the energy transition with hydrogen as an energy carrier and, on the other hand, the ambition to establish and expand a hydrogen infrastructure with the advertising of cross-industry service providers, with the identification of sustainable contributions and the resulting complementary competencies.

I met the established companies in the energy market and their initial rejection of hydrogen as an energy source and the fuel cell as an energy converter with meaningful results of the potential analyses, with the potential of the Power-to-X sector coupling, the identification of unique selling points and with the complex project impulses of the multi-layered interests of all those involved.
With perseverance and patience, even initial skeptics of hydrogen as an energy source and the fuel cell as an energy converter were successfully won over to a sustainable infrastructure in the energy transition.

With the EEZ Energy Dialogue, I am a long-standing member of the DWV German Hydrogen and Fuel Cell Association. The DWV is one of the successful mouthpieces in Europe for hydrogen as an energy source and for the fuel cell as an energy converter; the DWV speaks with the results of over one hundred industrial and research institutions.

I successfully contributed to the acquisition of members for the foundation of an advisory board for h2herten, from which the advisory board for the h2-netzwerk-ruhr emerged through expansion in 2008.

Picture above:
The hydrogen user center h2herten: the regional hydrogen nucleus in the northern Ruhr area.

Picture left:
14.06.2019: The opening of the hydrogen filling station at the h2herten user center; in the background the former coal mine Auf Ewald in the background.

Pictures: Dieter Mende; EEZ
Energie Energiewirtschaft Zukunftsenergien